I N MID-NOVE e-mails from Unit at the University of East Anglia involving some of the leading names in global warming research were made public, probably by a whistle-blower. The resulting scandal – dubbed "Climategate" by critics – was a public-relations disaster for the global warming industry. It seriously shook public confidence, not only in the scientists involved in the research but also in the lobbyists and politicians pushing for radical action to address a problem that might be more manufactured than real. The leaked e-mails revealed disturbing plans to either destroy or hide temperature data, manipulate the data to hide an observed decrease in temperature, and discussions of how to block the publications of global warming skeptics.

This behavior certainly does not help the United Nations' efforts to coordinate a new international global warming treaty to replace the Kyoto Protocol, which runs out in 2012. The primary goal of a U.N. meeting in Copenhagen in early December 2009 was to forge a new agreement for much larger reductions in greenhouse gas emissions than were targeted by the Kyoto treaty. While it looks unlikely at this writing that those legally binding targets will be agreed to in 2009, it is only a matter of time before another push for treaty negotiations and signing is made.

What follows is a brief overview of the science, economics and politics of anthropogenic global warming, with an emphasis on what is not being reported by the mainstream media. Given the multi-trillion dollar costs involved in greatly reducing greenhouse gas emissions, it is critical that the public understands what is at stake if we agree to legally binding limits, and just how poor the science is that is being used to justify these reductions.

THE PUTATIVE REASONS FOR REDUCING GREENHOUSE GAS EMISSIONS

The ostensible goal of any mechanism like the Kyoto Protocol is to reduce our consumption of carbon-based fuels: primarily coal, petroleum and natural gas. Burning of these fuels produces carbon dioxide, a so-called greenhouse gas. Greenhouse components in the atmosphere reduce the rate at which the Earth cools to outer space through infrared (IR) radiation. The most important greenhouse components are water vapor and clouds, with carbon dioxide representing a small – but not negligible – fraction of the Earth's total greenhouse effect.

The Earth's natural greenhouse effect acts somewhat like a lid on a pot of water that has been warmed on the stove. The lid reduces the rate at which heat is lost to the pot's surroundings and keeps the water warmer than if the lid were not there. Similarly, the greenhouse effect is a "radiative lid" that keeps the atmosphere warmer in the face of solar heating than if those greenhouse components of the atmosphere were not there.

It can be calculated theoretically that our addition of more CO_2 to the atmosphere has made the Earth's natural greenhouse lid about

1 percent more efficient at reducing energy loss to outer space, compared with preindustrial times. The expected result is a warming tendency, but the actual amount of warming is much more uncertain than you have probably been led to believe. While it is theoretically possible for more carbon dioxide to result in a future climate catastrophe, it is also possible that the amount of warming will be barely measurable in comparison to natural climate variability. The question is just how "sensitive" the climate system is, an issue I will return to later.

The most immediate and effective way to reduce greenhouse gas emissions would simply be to cease all economic activity, all commerce, in the world.

But for now, let us assume that global warming is caused entirely by humanity's use of fossil fuels, and that people and the biosphere are at great risk. How might we go about fixing the problem?

ENERGY IS REQUIRED FOR EVERYTHING WE DO

First, we need to understand that everything humans do requires energy, and the primary source of that energy today is fossil fuels. For instance, in 2005, about 86 percent of the energy consumed in the United States came from petroleum, coal and natural gas. About 8 percent came from nuclear power, and about 7 percent from renewable sources, primarily hydroelectricity.

The most immediate and effective way to reduce greenhouse gas emissions would simply be to cease all economic activity, all commerce, in the world. Since that is not a practical option, we can safely assume that societies will not voluntarily go down that road. Indeed, China and India are now going through their own Industrial Revolution, with rapidly expanding economies and fossil-fuel use. With hundreds of millions of their citizens still waiting to get electricity and clean water, it is unreasonable to expect that these huge developing markets for energy would just evaporate.

The other option is to use alternative forms of energy that produce little or no greenhouse gases. The goal seems simple enough. We've put people on the moon and sent robots to Mars, so you would think that clean energy would be a slam dunk.

In little more than 100 years, mankind has solved all kinds of problems through various technologies. Diseases have been prevented and cured, airplanes invented, computers, TVs, electronic communications and the Internet have been developed, and various pollutants have been reduced or removed altogether from consumer products, vehicle emissions and smokestacks.

Developing clean alternative sources of energy would seem to be just one more engineering problem to be solved. After all, there is free sunlight and wind energy all around us. If we can just tap into these renewable, inexhaustible sources of energy, we can all "go green."

Unfortunately, not all technological problems are equally easy to solve.

Since the ultimate source of most kinds of energy on Earth is the sun, you can think of fossil fuels as solar energy that has accumulated over very long periods of time. They are stored forms of concentrated energy, available even when the sun does not shine and the wind does not blow.

In contrast, solar and wind are more diffuse and intermittent forms of energy, requiring huge tracts of land to generate the same amount of energy as a single coal-fired or nuclear power plant. Solar panels gather that energy in real time, before it has had a chance to accumulate. Therefore, it's not just a matter of the existence of solar or wind energy in the environment; it is an issue of just how much of that energy can be usefully extracted on a continuous basis. Solar collectors in Earth's orbit would help by continuously providing two or three times as much energy as a land-based system of the same size. Unfortunately, the cost would be astronomical.

Hydroelectric power is attractive, since it has essentially no greenhouse gas emissions. Environmental concerns, though, are not only blocking the construction of new hydroelectric dams; they are leading to the removal of some of the existing ones.

Nuclear power is a particularly compact form of stored energy, but the disposal of spent fuel and safety concerns continue to prevent widespread acceptance of nuclear energy, especially in the United States. This source probably has the greatest potential for generating substantial quantities of energy to replace fossil fuels in the coming decades, but the public would need to become more supportive. The permit process and construction would take many years before a large fraction of coal-fired plants could be replaced.

Nevertheless, it is true that the world's supply of fossil fuels is gradually being depleted, and it is probable that renewable sources of energy will be increasingly relied upon in the future. But are we at the point today where a new green-energy economy can be encouraged to form and flourish? Is the transition to renewable energy just a matter of breaking old, dirty habits and forming new, healthier ones?

FORCING A REDUCTION IN FOSSIL-FUEL USE

Apparently, the free market is not leading to alternative forms of energy fast enough to address what Al Gore calls our "climate crisis." If the climate system really is as sensitive to our greenhouse gas emissions as the U.N., NASA's James Hansen and Al Gore claim it could be – an issue I will address later – then we need to drastically reduce our emissions as fast as possible.

And since energy use in the U.S. is a dwindling fraction of total global energy use, international agreements are necessary to provide any hope of stopping the growth of global emissions. This is why various command and control mechanisms are being discussed, under the direction of the U.N., in order to force reductions in carbon-dioxide emissions by governmental fiat.

Ignoring for the moment the need for international cooperation, how might we drastically reduce emissions at home? While one mechanism would be to mandate greater efficiency in energy use, it is important to recognize that a free-market economy already encourages energy efficiency. In fact, a free market is all about efficiency: efficiency of scale, of energy and raw material use, of labor, of balancing supply and demand, etc. This is why the amount of energy required to produce $1 of gross domestic product in the United States continues to fall year after year. Competition breeds efficiency.

The main argument one hears for greater energy efficiency is that a lot of energy is wasted every day. But as any engineer or scientist can tell you, just because a lot is wasted does not mean that the wasted energy could be easily reduced or recovered. Nevertheless, there are still some efficiency gains that can be made. For instance, new laws can require better insulation of buildings or increased fuel economy standards for new cars. But the up-front cost of making existing infrastructure and products significantly more efficient is very high.

There is little doubt that energy-efficiency gains will have some impact, but energy savings from these efforts will be small compared with the total energy needs of the world. The rate at which China and India are growing their economies would wipe out any energy savings through improved efficiency in the U.S. in a matter of weeks. All we would be doing is briefly delaying the inevitable.

If global warming really is caused by humans, then what we need are drastic reductions in our production of carbon dioxide in the coming decades – say, at least 50 percent reductions by 2050. This is why governments around the world are working on mechanisms that would reduce fossil-fuel use much more drastically.

There are a few ways in which this might be done. Since the EPA ruled on December 7, 2009 that carbon dioxide and other greenhouse gases are a significant threat to human health, they could start regulating any stationary sources of pollution exceeding 250 tons per year under the Clean Air Act. But if the EPA decides to go ahead with any such regulations on the larger emitters, lawsuits will tie the issue up in the courts for years to come.

A tax on carbon dioxide emissions is a fairly direct mechanism, since it would increase the cost of using fossil fuels and make alternative forms of energy more cost-competitive. It is the option most favored by environmentalists because it is the option most likely to succeed. The larger the tax, the less carbon-based energy will be used.

But a substantial carbon tax would be strongly resisted by a public that already feels overtaxed. So, a more indirect way to accomplish carbon emission reductions is to legally limit (cap) the amount of greenhouse gases that can be emitted by various industries and companies by distributing or selling a limited number of carbon allowances or permits. Then, the companies that are more successful at achieving greater reductions in carbon emissions could sell their leftover permits to other companies.

Such a "cap-and-trade" scheme is much less obvious than a tax, even though the goal is the same: punishing the use of fossil fuels. A cap-and-trade program is supported by the Obama administration and has already been adopted in the European Union, where it is called the Emissions Trading System (ETS). And now it is part of an energy bill being considered by the U.S. Senate that already passed in the House of Representatives.

In case you had not noticed, cap-and-trade is not a very popular option for reducing carbon emissions – except among politicians and those who will make money in the new system of carbon accounting, verification, buying and selling. While cap-and-trade is cloaked in free-market language by politicians, in a truly free market, the amount of energy a company uses would not be rationed in the first place. The market simply helps to spread the economic damage around more equitably.

Cap-and-trade will encourage all kinds of dubious financial transactions. It will be Enron on steroids. Cap-and-trade would set up a large new bureaucracy with various carbon-

---

Cap-and-trade will encourage all kinds of dubious financial transactions. It will be Enron on steroids.

---

trading and accounting mechanisms. The opportunities for cheating, favoritism and playing the system under cap-and-trade have many corporate interests, including Wall Street, drooling with anticipation. Partly for these reasons, many leading environmentalists oppose cap-and-trade.

In the process, the government also hopes to collect abundant new revenues by selling carbon indulgences – er, I mean allowances. I am told that the Obama administration has directed congressional Democrats behind the scenes to pay for any new health-care legislation with revenues generated by cap-and-trade legislation.

Of course, the extra money required for all of this will come from your pocketbook. Cap-and-trade is simply a more inefficient and obscure way to accomplish the same end as a carbon tax: making fossil fuels more expensive. It must be remembered that energy is required for everything we do, not just for transportation, but for the provision of goods and services of every kind. Increased costs from either a carbon tax or from cap-and-trade will simply be passed to the consumer.

"Big Business" cannot absorb the extra cost, since they would be forced out of business, along with the jobs, goods and services that they provide. Or they would have to relocate to another country that does not have similar environmental restrictions – in much the same way that California is increasingly outsourcing its polluting industries to other states in order to make itself appear green.

The Putative Reasons for the Green Jobs Illusion

The development of green jobs in a new green economy is an illusion. Sure, the government can indirectly force jobs out of one sector of the economy and into another sector, but the nature of those jobs makes all the difference in the world. For instance, we could have full employment today, with everyone working at green jobs, if we really wanted to. Half of us could dig holes in the ground, and the other half could fill the holes up again. How's that for a green jobs program?

The point is that what those jobs produce in terms of goods and services is more important than the mere existence of the jobs. And if the production of green energy requires many more jobs per kilowatt-hour of energy produced than, say, coal-fired or nuclear power, then the whole economy suffers. In fact, that would be a decrease in efficiency – exactly the opposite of what so many desire.

If green energy is economically competitive, then there is no need for legislation, regulation or subsidies. It will arise and grow organically. But the fact is this: Not only is green energy still relatively expensive, it cannot meet a substantial fraction of our energy needs. Even if we wanted to spend the extra wealth it will take to provide more green power, it will be very difficult from a practical standpoint to build and install enough wind- and solar-powered facilities to replace more than a small fraction of the electricity we currently get from coal and natural gas.

And until some method is devised for efficiently storing and recovering that energy, we will still require fossil-fuel power to back up solar power, when the sun does not shine, and wind power, when the wind does not blow.

It is unfortunate that even some petroleum companies have started to pander to public misconceptions in TV commercials, boasting a "diverse energy portfolio." But the truth is that green energy – at least on a large scale – is still a pie-in-the-sky illusion. Companies want to be viewed as green today mostly because it is an effective marketing strategy in a society that is obsessed with Saving the Earth.

THE ALTERNATIVE TO LEGISLATION

Yes, fossil-fuel supplies are finite and slowly dwindling. But history has shown that the best solution to problems of supply is to let the free market work. Shrinking fossil-fuel supplies and the increasing expense of reaching what is left in the ground will inevitably result in rising prices.

And this is exactly the signal the market

History has shown that the best solution to problems of supply is to let the free market work.

---

needs to spur the development of alternatives. New technologies will not occur overnight, of course. But everyone and everything we do needs energy, so it would be hard to find a better example of a huge, guaranteed demand waiting for new forms of supply.

Research that is already in progress in the private sector will eventually lead to new energy technologies that are both cost-competitive and deployable on a large scale. The universal – and growing – need for energy ensures that technological progress will be made without any intervention by the government and all of the inefficiencies that government intervention inevitably brings.

Those technological advancements cannot be simply legislated into existence by Congress or demanded by the president. In fact, any policy action that hurts the economy now might actually delay the development of alternative forms of energy. This is because the extra wealth generated by a vibrant economy is necessary in order to fund research and development efforts. When there is an economic downturn, R & D is the first place where companies scale back their investments.

Technological advancements cannot be simply legislated into existence by Congress or demanded by the president.

THE DIRTY LITTLE SECRETS OF GLOBAL WARMING SCIENCE

We have all heard – especially from Al Gore – that the scientific debate is over. It is claimed that "real" climate scientists are unanimous in their view that recent warming of the climate system is mostly, if not totally, the result of human activities. After all, CO_2 is a greenhouse gas, the Earth's natural greenhouse effect keeps the surface of the Earth habitably warm, and more CO_2 can be expected to cause some warming. Many people (including scientists) have simply assumed that any human impact on the climate system should be avoided, even if it is at great cost to humanity.

Even in his wildest dreams, Al Gore could not have imagined how successful his push to limit greenhouse gas emissions would be. Gore's movie *An Inconvenient Truth* used Hollywood visual and storytelling techniques to establish a supposed scientific connection between weather events like Hurricane Katrina and the tailpipe emissions from your SUV.

That movie did a masterful job of connecting in the viewer's mind dramatic weather events that happen naturally – tornadoes, hurricanes, floods, droughts, etc. – to human activities. For instance, large chunks of ice calving off of glaciers and plunging into the ocean were supposed evidence of anthropogenic (human-caused) warming. What Gore conveniently left out of his film, however, was the fact that ice breaking off of the Greenland and Antarctic ice sheets has always occurred and will continue to occur as long as snow keeps falling on them. Glaciers and ice sheets are dynamic systems, constantly on the move in response to the inescapable force of gravity.

Destructive hurricanes are another perfectly normal – if infrequent – example of nature's power. It has long been known that New Orleans is particularly susceptible to storm-surge flooding from a major hurricane, and it was only a matter of time before a hurricane-induced catastrophe occurred there. While it is true that hurricane damage has increased in recent decades, it is well-known that this is only because people continue to flock to our coastlines, where they build homes, businesses and infrastructure.

In fact, recently published evidence has shown that, after correction for our improved methods for finding and measuring the strength of tropical cyclones over the years, there has been no long-term upward trend in hurricane activity. All previous studies that claimed to find an increase in hurricane activity in recent decades were contaminated by a historical record that has a built-in bias toward fewer storms early in the record. With our global network of environmental satellites, it is no longer possible for a hurricane to form in some remote location without being detected. This has given the illusion of an increase in storm activity over time.

Ditto for tornado activity in the United States. While the continuing migration of people into the far corners of our nation, as well as the recent proliferation of video cameras, has resulted in an increase in the number of weak tornadoes reported since monitoring began in the 1950s, the number of strong tornado reports has not increased. This is probably because strong tornadoes leave considerable damage in their wake, which is hard to miss. There has likely not been any real trend in tornado activity, and the illusion of one exists for the same reason that a supposed increase in hurricanes has occurred: Our ability to detect these events has improved greatly over the years.

Yet there is much more convincing evidence that the global average temperature is warmer now than it was 30 or 50 years ago.

The thermometer-measured warming might have been exaggerated by spurious warming from land-use changes in the vicinity of the thermometers, but it is still very likely that

---

There is a widespread, quasi-religious assumption that nature was in a delicate state of balance before it was upset by the activities of humans.

---

a significant temperature rise has indeed occurred. The United Nations Intergovernmental Panel on Climate Change (IPCC) even claims that warming is "unequivocal."

But just how good is the scientific evidence that this warming was caused by human activity? Indeed, even if all kinds of severe weather have increased with warming, what is the proof that all of this was caused by human activities? Are there alternative explanations for recent warming that would relegate the impact of humans to the level of noise? Have scientists adequately ruled out other possible explanations for recent warming, other than the greenhouse gas emissions from our use of fossil fuels?

The fact is that scientists and politicians alike have been too quick to assume causation when it comes to climate change. Like ancient tribes of people who made sacrifices to the gods of nature, modern societies are too easily convinced that they are the ones responsible for weather changes and that they can somehow atone for their environmental sins through changes in behavior. It is fairly easy to demonstrate the religious nature of our belief that humans now control the climate system.

THE RELIGIOUS NATURE OF CLIMATE CHANGE BELIEFS

Imagine if the activities of humanity were actually destroying carbon dioxide, rather than creating more of it. There would be howls of protest that we are destroying life on Earth.

After all, carbon dioxide is essential for photosynthesis on land and in the ocean, so destroying the stuff would be an obvious assault against life on Earth. How can it be that adding *more* CO_2 to the atmosphere is a danger to life on Earth, too?

The answer is that there is a widespread, quasi-religious assumption among laypeople and scientists that nature was in a delicate state of balance before it was upset by the activities of humans. As part of this belief system, the CO_2 content of the atmosphere that existed before we started altering it is assumed to be that which the Earth "prefers."

But this worldview has nothing to do with science. In fact, there are hundreds of published research studies that document the benefits of more atmospheric carbon dioxide on the growth of all kinds of plants. Large greenhouses pump in three times the atmospheric concentration of CO_2 to encourage faster plant growth. Some recent evidence is suggesting that more carbon dioxide also will benefit marine life in the oceans and that fears of "ocean acidification" could be largely unfounded.

The assumption that everything humans do necessarily hurts the environment is a powerfully seductive belief for many Westerners who harbor angst over the myriad and amazing ways humans have found to rearrange the raw materials of the Earth to suit our needs. Our cars and planes and machines and gadgets and conveniences are all very nice, but they also complicate our lives. We sometimes yearn for simpler times, a bygone era when life was less hurried.

Yet that way of life has not been relegated to the dustbin of history. It is still lived by more than 1 billion people on the Earth today who have no electricity, no refrigeration, no clean water and who must gather wood and animal dung to use for heating and cooking. Life for these people is dirty, smelly, difficult – and short. Given the chance, most of them would gladly change places with you.

I contend that this belief system, combined with political and financial aspirations of politicians and government managers, has helped lead to a strongly biased body of scientific research. It is not so much that previous global warming research has been bad, but that it has been too narrowly focused on mankind and too incomplete in its investigation of nature as a potential source of climate change. In other words, climate researchers have stopped developing and testing alternative hypotheses, a process that is necessary for the advancement of science.

What Scientific Consensus?

The IPCC was formed more than 20 years ago to build the scientific case for humans as the cause of global warming – not to find alternative explanations for it. The IPCC is the ultimate source of what is claimed to be the scientific consensus on global warming.

The IPCC depends upon many of the world's climate scientists to survey what is known about climate change and global warming. Those scientists have done a reasonably good job of summarizing what is known about how the climate system works.

But the politicians, bureaucrats and a handful of activist scientists have decided what all of that complex science means for humanity – and the human behaviors that need to be controlled. The peer-reviewed and published science on the subject of anthropogenic global warming is much more equivocal than the IPCC leadership claims. The IPCC indeed rests its case on the peer-reviewed scientific literature, but they extrapolate well beyond what the science can actually support.

So exactly what is this consensus that we keep hearing about? This is never explicitly stated. Is it that "global warming" has occurred? No. Even a skeptic like me would agree that there has been significant warming in the past 30 to 50 years. In that sense, there is no such thing as a "global warming denier." I do not know of any global warming skeptic who denies that warming has occurred in recent decades.

And the pejorative "climate change denier" is particularly ironic, since it is the global warming alarmists who deny that natural climate change exists.

Contrary to what you might have been led to believe, politicians and savvy governmental representatives, not the scientists, have taken it upon themselves to decide what the consensus is. And when you hear of scientific societies making official statements on global warming, they are merely parroting what they have heard. The experts who supposedly all agree on global warming are just assuming that someone else who is more expert than them has all of the evidence and answers. In the end, the supposed consensus is never explicitly defined. This allows anyone invoking it to decide what it means, or anyone hearing it to imagine they know what it means.

THE MISSING SCIENCE OF GLOBAL WARMING

Even the layperson understands that scientific truth (if such a thing exists) is not determined by majority opinion. Albert Einstein once said that no amount of research could ever prove his theories to be correct, but it would take only one experiment to disprove them. The same is true of global warming. It would take only one research study to demonstrate that global warming is more natural than man-made and that future warming in response to our greenhouse gas emissions will be small.

Fortunately for the IPCC leadership, though, that is unlikely to happen any time soon. The reason is that we do not have sufficiently detailed and accurate global observations of the climate system over a long enough period of time to understand the role of Mother Nature in causing climate change. You cannot study that which you do not have the data to study. For instance, the most recent period of warming that started in the late 1970s might have been mostly caused by a 1 percent or 2 percent decrease in global average cloudiness. Sufficiently accurate global satellite data to document such a change, however, has only existed since about 2000.

In contrast, we have very good records of

the CO_2 content of the atmosphere for at least the past 50 years, so scientists have simply assumed that increasing carbon dioxide is to blame. If that is indeed the case, then the warming caused by such a weak influence (carbon dioxide) means that the climate system is quite sensitive, and so substantial warming will continue as we continue to burn fossil fuels.

But if it turns out that past warming is mostly natural in origin, then it takes a much stronger push to cause warming, which indicates an insensitive climate system. In that case, the natural warming can be expected to end at some point, and little additional warming from our greenhouse gas emissions can be expected. Since it has not warmed in the past 10 years, it could be that the natural warming has already ended.

So, while it is true that the IPCC has ruled out the most obvious external influences on the climate system – changes in sunlight intensity, ozone depletion and volcanic eruptions – it has essentially ignored natural sources of climate change generated by the climate system itself. The IPCC can then truthfully claim that there are no "known" mechanisms other than our greenhouse gas emissions, which are sufficient to explain recent warming.

This leaves the impression with the public and policymakers that the IPCC has actually ruled out all other potential sources of warming based upon the evidence, when the truth is that it has ruled out only the few it has sufficient data to study. Other potential sources are just ignored. And since the science of climate change is so complex, the IPCC does not even need to mention the omission. I suspect that most climate experts are not even aware that the issue exists.

It might be difficult to believe, but what turns out to be the most commonly held public belief about global warming – that it is just part of a natural cycle in the climate system – has seen virtually no scientific research. And it is this explanation that I believe will eventually be proved correct.

* * *

Before global warming became the most popular climate research topic, most of the evidence that had been published over the years suggested that periods of natural global warming and cooling are the rule, not the exception. A number of indirect measures of past temperature (temperature "proxies") suggest that it might well have been warmer during some decades around 1000 A.D., during the Medieval Warm Period, than it has been recently. Our present warmth could well be the culmination of a natural process that began hundreds of years ago, at the end of the Little Ice Age. Since most of that warming – including retreating glaciers – occurred well before fossil-fuel use could be blamed, there are indications that there are natural climate variations at work that are, as yet, not understood.

This is why the IPCC so strongly pushed the so-called "hockey stick" reconstruction of temperatures over the past 1,000 years, which essentially did away with the Medieval Warm

Period and Little Ice Age. In their place was a long period of little temperature change, ending with rapid warming in the 20th century, giving a hockey-stick shape to the temperature reconstruction.

But as revealed by a National Academies review panel, it turns out that the hockey-stick shape was essentially forced into the data analysis. This appears to be a not-so-subtle attempt to make it look like climate change did not exist before the Industrial Revolution. As a result of the review, the IPCC no longer highlights the hockey stick. Recently uncovered private e-mails revealed on November 19, 2009, from the U.K.'s Climatic Research Unit suggest that even the original developer of the hockey-stick data set had doubts about his results. Those e-mails also reveal collusion to interfere with the peer review process in order to prevent scientists with opposing views from getting their research published.

Claims that our present warmth is somehow unprecedented are simply opinions, educated guesses that can be neither proved nor disproved. And this happens to be very convenient for those who are intent on regulating global energy use.

THE PUSH TO ACT NOW

To the extent that it actually exists, anthropogenic global warming is a slow process. Even if we assume that the IPCC's predictions of a sensitive climate system are correct,

---

Any regulatory or legislative action that the Obama administration or Congress advocates will be all pain for no measurable climate gain.

---

the average future warming rate is expected to be little more than two hundredths of a degree Celsius per year. If the minority view that the climate system is insensitive is correct, then that warming would be minuscule – thousandths of a degree per year. At that level, we need to be much more concerned about natural climate change than anthropogenic climate change.

So why is it that there is such a strong push to act now? Why do politicians keep claiming we have only a few years left to save ourselves, or that the latest U.N. meeting is (once again) our last chance to avert global disaster? Furthermore, why is it that any published research about the climate system possibly being much less sensitive than the IPCC claims it to be is greeted with scorn rather than relief? Why does the news media refuse to report on such research when it is published?

And finally, why do IPCC scientists refuse to debate scientists like me? Is it because they know their case is much weaker than what is portrayed by politicians and popular culture? These are questions you need to ponder when you are asked to support global warming legislation.

The bottom line is that any regulatory or

legislative action that the Obama administration or Congress advocates will be all pain for no measurable climate gain. It will destroy wealth, increase the price of virtually everything, drive industries overseas where they can pollute even more, substantially grow government and greatly restrict personal freedoms.

There are as of yet no large-scale replacements for fossil fuels. If you want a 50 percent or greater reduction in carbon dioxide emissions, you are going to have to reduce the use of fossil fuels by 50 percent. In order to accomplish that on a global basis, we will need new energy technologies that do not yet exist.

These new technologies cannot be simply legislated into existence. History has shown that it is instead much better to let the free market handle the solution to this problem. And since there will always be a demand for energy, you can be sure the solutions will come.